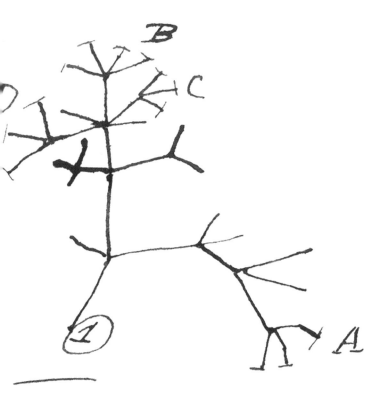

Darwin's tree of life sketch from his notebook on transmutation of species.

Series 117

This is a Ladybird Expert book, one of a series of titles for an adult readership. Written by some of the leading lights and outstanding communicators in their fields and published by one of the most trusted and well-loved names in books, the Ladybird Expert series provides clear, accessible and authoritative introductions, informed by expert opinion, to key subjects drawn from science, history and culture.

The Publisher would like to thank the following for the illustrative references for this book:
Page 13 from images © (Theodor) American Philosophical Society/Science Photo Library and (Thomas) Imango/Getty Images; page 17 from photo © 12/Getty Images; page 19 from photo © A. Barrington Brown, Gonville and Caius College/Science Photo Library; page 21 from photo © Universal History Archive/Getty Images and © National Library of Medicine (NLM); page 31 from photo © Associated Press/Shutterstock.

Every effort has been made to ensure images are correctly attributed; however, if any omission or error has been made, please notify the Publisher for correction in future editions.

MICHAEL JOSEPH

UK | USA | Canada | Ireland | Australia
India | New Zealand | South Africa

Michael Joseph is part of the Penguin Random House group of companies whose addresses can be found at global.penguinrandomhouse.com

First published 2018
001

Text copyright © Adam Rutherford, 2018

All images copyright © Ladybird Books Ltd, 2018

The moral right of the author has been asserted

Printed in Italy by L.E.G.O. S.p.A.

A CIP catalogue record for this book is available from the British Library

ISBN: 978–0–718–18827–6

www.greenpenguin.co.uk

Penguin Random House is committed to a sustainable future for our business, our readers and our planet. This book is made from Forest Stewardship Council® certified paper.

Genetics

Adam Rutherford

with illustrations by
Ruth Palmer

Ladybird Books Ltd, London

Introduction

Genetics is the study of inheritance. For us, and other animals, that means it's the study of sex, families and disease. Most living things on Earth are not us, and don't have sex, but the genetics of those sexless creatures is largely similar. A simple code spells out a vocabulary that is universal in all living organisms. It is a code inherited from our parents, whether they are a male and a female animal, male and female parts of the same organism, or simply a single cell, such as a bacterium, that has split in two. The code is written in the molecule DNA, beautiful in its structure, and similarly elegant in how it works.

It doesn't take a geneticist to know that individual creatures within a species are more similar to each other than to other organisms, and that children resemble their parents more than random strangers. This is the central question of the field of genetics: how does inheritance work?

Genetics is also the study of DNA, a science only a hundred years old, in any real sense, only fifty years old with any detail. In that short period, our knowledge of DNA has upturned every aspect of biology, spawned multi-billion-dollar industries, rewritten evolution and transformed medicine. There are no areas in the science of life that have not been irreversibly changed by our decryption of this simple code.

Life on Earth all sits on one epic, sprawling, incalculably tortuous family tree. Everything on that tree is coded in DNA. This is the idea that would eventually tie together the biggest concepts in science. But only after a faltering beginning.

A false start

Though DNA would be discovered properly in the twentieth century, it was first identified in 1871. During the Franco-Prussian War, in an age before antibiotics, soldiers were dying all over Europe from gangrenous amputations and seeping open wounds. In Tübingen in Germany, a young doctor called Friedrich Miescher was interested in the chemical makeup of human cells, and had the bright – if slightly ghoulish – idea of extracting cells from the abundant pus-soaked bandages from the legions of soldiers slowly rotting to death in the hospital next door. Miescher isolated and purified a new substance from the white blood cells in pus. He analysed the chemistry of this extract, and noted that it was rich with phosphate and contained no sulphur. No other bodily extract had been discovered with such high amounts of phosphate, and proteins often have sulphur in them. Because he only found it in the central nucleus of the cells, he named it 'nuclein'.

Miescher continued to work on nuclein for a few more years, later extracting it not from white blood cells, but from salmon sperm. The samples still exist in the University of Tübingen, a greying powder in small, dark-brown glass bottles. Modern tests have shown that Miescher's nuclein is indeed DNA – which is rich in phosphate but contains no sulphur. But this grisly discovery would play no further role in the revolution that was about to begin.

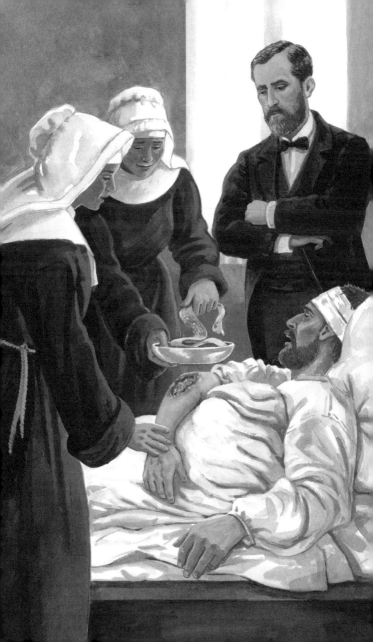

Peas, please

At about the same time, a bit further to the east, the Moravian scientist Gregor Mendel was beavering away in his garden laboratory. Mendel is invariably described as a monk, which he was; however, his contribution to science is colossal, whereas the legacy of his monkery remains of only passing interest.

In the monastery in Brno (pronounced *Brno*), he bred more than 29,000 pea plants together and looked at flower colour and pea wrinkliness, and a few other traits. What he saw was that individual characteristics were passed down independently of others, and that the mixing of two traits, such as purple and white flowers, did not result in a blend (pinky flowers), but a ratio of purple and white flowers that followed a very specific pattern.

Both of these observations form the foundations of our entire knowledge of genetics and inheritance. The elements that coloured the flowers were discrete, and didn't merge in the next generation. He was unwittingly describing the gene – the basic unit of inheritance, though the name would come later, at the beginning of the twentieth century.

The best idea anyone ever had

At exactly the same time, a few hundred miles west of Mendel's monastic lab, Charles Darwin was beetling away, coming up with the concept of evolution by natural selection. He had observed all manner of beasts on his travels around the world on HMS *Beagle* in the 1830s, and at his home of Down House in Kent. He showed that life is four-dimensional: creatures were not fixed in time and could be changed via selective breeding. Daft-looking pigeons had been bred for competition over many generations, but they were still all rock pigeons, despite looking rather silly, and utterly different.

The question was how and why creatures had changed in nature. His answer was this: any characteristics in a species varied between individuals in a population – within a herd of deer, some have bigger antlers than others; within a population of any species, some are naturally better at resisting infection. These differences were a pivot on which nature could act. The more suitable a particular trait was for the local environment, the more it would be passed down the generations. Over time, traits, and ultimately species, would mutate into ones which were best adapted to the changing environment – descent with modification.

It was pretty much the best idea anyone ever had. Natural selection explained how life on Earth evolved. Darwin didn't know how the modification was being passed down to each generation, or how that information was responsible for the traits themselves. That would all come later, when the strands of Darwin's work, Mendel's peas and others would be united in genetics. Alas, Darwin died unaware of Mendel's research, which is a shame really, because between them they had laid the foundations of all biology.

Mostruoso!

At the beginning of the twentieth century, scientists all around the world were trying to work out precisely what was transmitting information from generation to generation.

Theodor Boveri, working in the Marine Station on the beach of the Bay of Naples, used animals abundant in those waters: sea urchins. Making an urchin ejaculate is alarmingly easy: you simply give it a good shake and then place it upside down in a glass of water. The semen then flows out. With these samples being so readily available, Boveri performed all sorts of experiments, and showed that chromosome number was crucial to good health. He described the deformed urchins with the wrong number of chromosomes as *mostruoso* – 'monstrous'.

Meanwhile, in New York, Thomas Hunt Morgan was breeding fruit flies by the bucketload, and was beginning to show that on specific chromosomes there were regions that seemed to correspond to the eye colour of his flies.

Genes – the units of inheritance – sit within chromosomes. Most organisms have two sets of chromosomes, one inherited from each of their parents. Humans have twenty-three pairs, though one of the pairs is not always a pair. Men have a Y chromosome inherited from their fathers and an X from their mothers, whereas women inherit an X from each parent.

Genes, evolution and inheritance

Genes are a part of chromosomes, and chromosomes are made of DNA.

In the first half of the twentieth century, scientists were inching forward to an understanding of the mechanism and the material of inheritance. In the 1930s, Oswald Avery and his team took DNA from a virulent strain of the bacteria that cause tuberculosis, and transferred it to bacteria that was innocuous. The bacteria thereby became disease-causing agent. They had shown that DNA was the hereditary material. It was what carried the crucial information from cell to cell.

Meanwhile, mathematicians had grabbed hold of Darwin's theory, and worked out the ways genes were passed through populations over time. The gene is the unit of inheritance, and it is the thing on which natural selection acts. Later, this became known as 'selfish gene' theory, popularized by Richard Dawkins. The measure of evolution is the way the frequency of genes in a population changes over time. If a gene for a bigger tail on a peacock helps the peacock mate more often, then that gene will spread through the generations more efficiently, and that gene will have survived into the future. It has been selected.

The best way for individual genes to achieve immortality is in concert with other genes, in coordination inside an organism. We are merely husks for our genes to succeed.

Franklin, Crick and Watson

Now that we knew that genes were made of DNA, the race to find the structure of DNA was on. By the 1950s, scientists knew its chemical composition, but not how it was built, which might explain how inheritance works.

Maurice Wilkins, Rosalind Franklin and her student Ray Gosling were experts at a technique called X-ray crystallography, which is used to work out the three-dimensional shape of a molecule. In 1952, in their basement laboratory at King's College London, Franklin and Gosling took a series of X-ray photos of DNA, and made precise measurements which would help reveal vital clues to DNA's structure.

At the beginning of 1953, some of their data was shared with an American scientist in Cambridge called James Watson and his British colleague, Francis Crick. They had been working on the structure of DNA, but had so far failed to find the answer. Using Franklin's data, they were able to succeed.

By the beginning of 1953 Franklin was winding up her work on DNA. The atmosphere at King's was toxic, and she was treated unpleasantly by some male colleagues. Later, she was sexistly derided by James Watson in his account of the discovery (though he was subsequently honest and more generous in his descriptions of her essential role in the discovery). She was a brilliant scientist, but died of cancer aged thirty-seven, before the Nobel Prize was given to Crick, Watson and Maurice Wilkins in 1962 – Nobels are not awarded to the dead. Nevertheless, her contribution to the story of genetics makes her one of the most important scientists in history.

The double helix

The shape of DNA was unveiled to the world on 25 April 1953, in a study authored by Crick and Watson. The final line may be the greatest understatement in scientific history:

It has not escaped our notice that the specific pairing we have postulated immediately suggests a possible copying mechanism for the genetic material.

Odile Crick, Francis's wife, drew the first double helix, a stylized version of a molecule that resembles a ladder twisting to the right. In life it is constantly unwinding and unzipping its two strands as it enacts its coded messages.

The spiral is inherent to its function and that is why solving its structure is one of the greatest scientific discoveries of the twentieth century. Those two strands are held together by four chemical letters – A, T, C and G – and these form pairs. G can only couple with C, and T can only couple with A. When you tear the two strands apart, the pairs are separated. But on each strand you have the necessary information to replace the missing strand. So if you rip a double helix in half down the middle, you can make two identical double helix molecules.

That's what happens every time a cell divides. All of the DNA in the cell is split in two, and duplicated. The two cells that remain contain exactly the same DNA as the parent. That's what has happened when you grow from a single fertilized egg in your mother's womb into a baby and then an adult, every time you cut yourself and heal, and in fact in every dividing cell for the last four billion years.

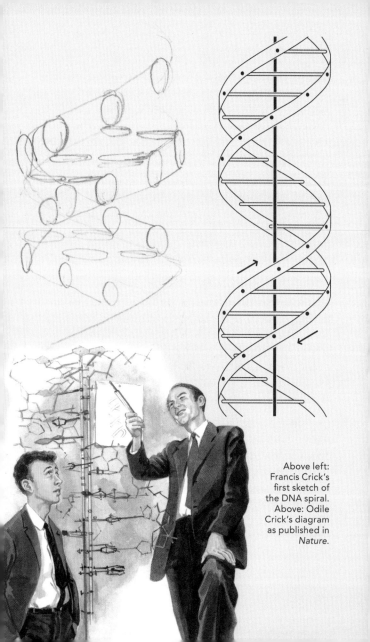

Above left: Francis Crick's first sketch of the DNA spiral. Above: Odile Crick's diagram as published in *Nature*.

The code

Crick and Watson had figured out the structure of DNA, and saw that this shape meant it could be copied over and over again. What they hadn't worked out was how DNA contained information, in the form of genes. The work of decoding DNA would be led by Crick and a host of other scientists over several years.

All life is made of, or by, proteins. Keratin in hair is a protein. The fibres in muscle cells are proteins. Bone is not a protein, but is made by proteins. All proteins themselves are long strings of molecules called amino acids. All living things use twenty-one amino acids to make all proteins, and the order of the amino acids determines what the protein does.

Two scientists in Bethesda in the US were the first to decipher a word in the language of genetics. Marshall Nirenberg and J. Heinrich Matthaei showed that an amino acid is coded by three specific letters in DNA, called a codon. Francis Crick saw Nirenberg present his findings in Moscow in 1961, and immediately knew that the first steps towards cracking the full code had been taken.

Over the next few years, the codons that encrypt all the other amino acids were discovered. There are sixty-four combinations of the four letters of DNA in codons of three. Each of these combinations corresponds to one of the twenty-one amino acids, and a few for the instruction for a protein to end, like a full stop.

Marshall Nirenberg with a molecular model.

	U	C	A	G	
U	Phe	Ser	Tyr	Cys	U
	Phe	Ser	Tyr	Cys	C
	Leu	Ser	STOP	STOP	A
	Leu	Ser	STOP	Trp	G
C	Leu	Pro	His	Arg	U
	Leu	Pro	His	Arg	C
	Leu	Pro	Gln	Arg	A
	Leu	Pro	Gln	Arg	G
A	Ile	Thr	Asn	Ser	U
	Ile	Thr	Asn	Ser	C
	Ile	Thr	Lys	Arg	A
	Met	Thr	Lys	Arg	G
G	Val	Ala	Asp	Gly	U
	Val	Ala	Asp	Gly	C
	Val	Ala	Glu	Gly	A
	Val	Ala	Glu	Gly	G

Codon chart.

Disease

We've known for all of history that diseases can be passed down through families. In the Jewish religious text the Talmud, there is a description of boys who are excused from circumcision if their brothers or cousins died after bleeding unstoppably. We now know that it was describing haemophilia.

We know that all cancers are genetic diseases, as tumours are caused by the unruly growth of cells resulting from mutations in genes that would normally keep cell division in check.

Some diseases follow very clear patterns of inheritance in families, rather like the pea characteristics in Mendel's experiments. Because the family trees of people with conditions such as cystic fibrosis or Huntington's disease follow these patterns, in the 1980s these conditions were the first to have their genetic basis described.

With the genetic code finally fully solved, the reason why diseases ran in families could be understood, and the genes that were at fault discovered. Cystic fibrosis is caused by a gene that is cut short and therefore makes a stunted protein which doesn't work properly in the lungs. Huntington's is caused by a gene with a section in the middle of it that has expanded more that it should have.

Over the next few years, the race was on to find genes for all inherited diseases.

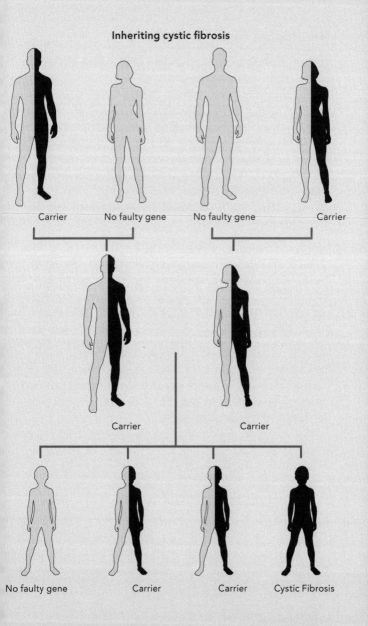

The Human Genome Project

Labs around the world were toiling away trying to find the genes responsible for diseases of interest. Progress was slow, and highly competitive. The Human Genome Project was set up as an attempt to speed things up. If genes were at the core of diseases and normal human variation, then surely a more efficient way of investigating them would be to sequence the whole of an average human genome – all our genes taken together – and create a database that researchers could search freely within, without having to identify each gene one at a time. So labs all around the world joined together in the 1990s to try to read the entire genome of a human being – three billion letters of DNA.

In 2000, President Clinton announced that the sequence of the human genome was complete, and the first draft was published six months later. In fact, it wasn't really complete, and it would be worked on for several more years. The project had cost $3bn.

Nowadays, we can sequence a complete genome for less than $1000, and millions of people have had their DNA sampled and read, some for health reasons, others just out of curiosity. When the first genome of the HGP was published, the results were very surprising.

Too few genes

There were two big surprises when the first human genome was published.

The first was that almost none of the human genome is made of genes. Of the three billion letters of DNA, less than 2 per cent is coded as genes. Some of the rest is scaffolding and structures that allow the formation of those nice neat chromosomes that we see during cell division. Some of it makes up instructions for genes – switches that say 'on' or 'off' to a nearby gene; after all, only a handful of the 20,000 genes in every cell need to be on at any one time.

Lots of it is huge sections that are repeated, and we don't really know what they are for, but we do know they're important. Some of it might not have any role at all – it might be old junk left over from evolution which once had a role but has rusted into obsolescence.

The second big surprise was that we don't have very many genes. We have fewer than a water flea, or a banana. Many scientists had once thought that there might be one gene for every characteristic – hair colour, eye colour or even ones that could cause disease if they were mutated. But there are not nearly enough genes for humans to be understood that way. In fact, we have only around 20,000 genes – about the same number of parts as an average car, including every screw.

These surprises meant that we had – and still have – a lot to learn about how our genomes work.

banana **36,000** genes

water flea **31,000** genes

human **20,000** genes

car **20,000** parts

Humans are quite complicated

Human genetics, it turned out, is much more complicated than many scientists anticipated, which might have seemed a little shortsighted given how complicated people are. We now know that some genes do many things. The effects of all genes are a complex interaction between DNA and the environment – so-called nature and nurture. We used to say 'nature versus nurture', but these two things are not in conflict. DNA – nature – enacts its code in cells, organs and bodies and in the world – nurture. A better phrase is 'nature via nurture'.

Though the first disease genes followed fairly straightforward patterns in families, we now know that almost all human characteristics involve many genes in complex concert with the environment. Versions of some genes can have an association with certain diseases or traits, but often the impact of them is relatively small. When it come to complex human behaviours such as intelligence, or complex diseases such as schizophrenia, dozens or even hundreds of genes appear to be involved, and these only account for a proportion of the probability of a person having that disease or trait.

Sometimes headlines say that scientists have discovered the gene 'for' a disease, or a set of human behaviours. This is almost always not true. There are no genes 'for' any particular human trait, only ones associated with a probability of that trait. Genes are not destiny.

But at the birth of genetics that idea was firmly rooted, and led to some of the most heinous crimes in history.

A murky past: eugenics

Genetics has a politically dark past. Many of the techniques and statistics that we still rely on today were invented by a Victorian scientist called Francis Galton. His legacy includes publishing the first weather map (actually printed the day after, so arguably of limited use), the basis of fingerprint analysis, and a vented hat to cool the head while thinking hard (though it might have been easier to simply take the hat in question off).

Galton was Charles Darwin's half-cousin, and somewhat enamoured of Darwin's fame. He thought that genius ran in families, and that humans segregated naturally according to race. He began collecting data on people, in order to work out the biological basis of the differences between people.

In doing so, he effectively became the father of human genetics. Many people thought at the time that Galton's ideas about improving the 'stock' of people – in the parlance of the day – was desirable for society, notably to find fitter men to fight in colonial wars. This was the birth of eugenics, a word that Galton himself coined.

Over the decades that followed his death, it wasn't just the Nazis who embraced eugenics. Churchill, Roosevelt, Marie Stopes (who wrote love letters to Hitler) and many public figures at all points on the political spectrum thought eugenics was a good idea. Many countries enacted eugenics policies to reduce the number of 'undesirables' in the population, including drunks, homosexuals and people with mental-health problems. In the United States, tens of thousands of people were involuntarily sterilized throughout the twentieth century, and even into the twenty-first.

Race

Francis Galton was a racist. He wrote that:

negroes possess too little intellect, self-reliance, and self-control to make it possible for them to sustain the burden of any respectable form of civilization without a large measure of external guidance and support.

Although there is natural human variation, and this change is broadly geographical, there are no sets of genes that correspond exclusively with a group of people that we might describe as a race. It's not even possible to say how many races there are.

Evolution's deception is that things like skin colour and hair make groups of people look superficially similar, while overall differences in DNA are not visible. Though superficially, we might cluster darker-skinned Africans together because of skin colour, their genomes are more different to each other than to most people's outside of Africa.

Diseases historically labelled as specific to races turn out not to be exclusive at all. Sickle cell anemia is not a black disease as is often said, but is common in people who have evolved in areas with endemic malaria. Tay–Sachs disease was identified in the nineteenth century as a condition specific to Jews, but it is also common in Cajuns and French Canadians.

We now know that the way we talk about race does not correspond with what genetics tells us about human variation. The irony is that Galton wanted to show that people could be categorized according to crude racial definitions, and his ultimate legacy was the opposite. Genetics has shown that race is a concept that is not scientifically useful or valid.

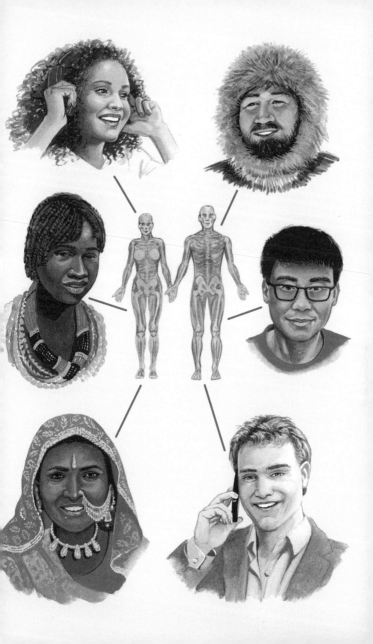

Evolving genes

The Human Genome Project allowed us to compare the DNA of people all around the world to help us understand human variation. Scientists also turned to sequencing the genomes of whatever they could get their hands on. Reading DNA became cheaper and easier, and all manner of creatures joined an exclusive list – the Genome Club. Rats, bees, chimps, wheat, mice, fruit flies were among the first to join the menagerie, and all showed fascinating similarities in their DNA. The genes to make an eye are virtually the same in all creatures with eyes – and this could be demonstrated by switching eye genes from a fly and eye genes from a mouse: both animals still made their own eyes, even though the genes were from a different species. Genes that make brains or nerve cells or legs are very similar in all creatures with brains, nerves or legs. Lots of the genes are effectively the same in all organisms great and small, from the blue whale to a bacterium.

These discoveries again showed Darwin's theory of natural selection to be spot on. The fact that the genes were the same in very different creatures indicated common ancestry. We could even use DNA to work out how long ago two species split from each other, by calculating how different their genes were. The more similar genes from two different living species are, the more closely related they are on the evolutionary tree. And in the twenty-first century we aren't even limited to looking at DNA from living creatures.

Very similar genes make eyes in all sorts of creatures.

Old bones

Since the 1980s, we've been picking up traces of DNA at crime scenes and using it to identify perpetrators or victims of crime. But it wasn't long before the Genome Club began admitting humans who had been dead for tens of thousands of years. In the right conditions, DNA is very stable, and in 2009 scientists managed to extract and read the whole genome of a Neanderthal man – a species of human that had become extinct more than 35,000 years ago. Within a couple of years, a genome was extracted from a single tooth and a finger bone found in a Siberian cave, and turned out to be another extinct type of human, called the Denisovans (after a hermit called Denis, who lived in the cave in the eighteenth century).

These two genomes showed that modern humans carry DNA from both Neanderthals and Denisovans, and possibly another human species that we haven't yet identified. Our ancestors had sex with these other human species many times, and we carry their DNA to this day. These new genomes have helped us redraw the map of human evolution, and of human migration over the last half a million years.

Scientists have become so good at detecting DNA that we can now even extract it from the soil in caves that once had humans in them. They might have been where someone died and all that is left of their mortal remains is DNA. Or it might be that humans used a cave for a huge communal toilet. Either way, it means we can genetically profile ancient humans without even needing their bones.

Genetic genealogy

Your family tree probably peters out after a few generations up from you, because people in history are difficult to identify. Legal papers and documents become more and more scarce as we go further back in time.

Now that gene sequencing has become cheap and quick, lots of companies have sprung up that offer to reveal your ancestry using your DNA. For around £100, you spit in a tube and send it off. A few weeks later you get a breakdown of where DNA similar to your own can be found on Earth today. That doesn't mean where your DNA came from in the past – there is no method for doing that – only where people who share your DNA can be found nowadays.

This information is fun, but isn't necessarily very informative. DNA is very useful in identifying very close family members, and reports of it being used to find unknown cousins or parents of children who were adopted are becoming more and more common. But when it comes to deep family relations, DNA is not a very effective tool. Some companies might tell you that you are descended from Vikings, or from Saracens, or obscure Anglo-Saxon noblemen. This may be true, but only because it is true for virtually everyone.

Your Majesty

Everyone loves to discover that they are descended from some illustrious lineage. In truth, we are all descended from royalty. There is an almost 100 per cent chance that if you are light-skinned and British, you are directly descended from Edward III. And all Europeans are descended directly from the ninth-century ruler King Charlemagne. Or if you are broadly East Asian you are descended from Genghis Khan. Not everyone can prove it using traditional genealogy, but with genetics and statistics it is a certainty.

You have two parents and four grandparents, and with each generation up, that number doubles. If you keep doubling, by the time you get to the tenth century, you will have more than one trillion ancestors on your family tree, which is thousands of times more people than have ever existed.

Family trees only look like trees for a few generations, and then they begin to collapse and form huge nets, meaning that you are descended from the same people many times over. In fact, by the time you get back to Charlemagne, all lines of all family trees cross through all people who existed at that time. That means that anyone in the tenth century with living descendants is an ancestor of everyone today. If you are broadly European, you are literally directly descended from Charlemagne.

If we go back to around 3,400 years ago, all humans from that time who have living descendants are the ancestors of all humans alive today.

Origin of life

We can go back even further. DNA takes us all the way back to the root of Earth's family tree. Every living thing we have ever found has DNA in its cells, and uses the same coding, and the same proteins, and the same basic metabolism. This points to the idea that there was a single origin of life on Earth.

We call this entity Luca – the Last Universal Common Ancestor. Luca wasn't the first life form, but was the root of the tree of life that all living things are on. We think Luca lived around 3.9 billion years ago, and was probably a thing a bit like a cell that lived in rock in a hydrothermal vent at the bottom of the sea, and had DNA and genes.

At the base of the tree of life it doesn't look much like a tree, because for a billion years or so the first single-celled organisms – bacteria and archaea – swapped genes with each other, just like they do today. It's more like a tangled web than a tree. We share genes with Luca, and by comparing over six million DNA sequences, scientists have worked out that Luca had over 350 genes.

Around two billion years later, one cell crawled inside another, and eventually was co-opted to form mitochondria – tiny energy generators that we and all complex life use to this day to power our cells. Acquiring those mitochondrial genes enabled life to become multicellular and complex, from plants to mushrooms to seahorses and ultimately to us.

Darwin sketching his first evolutionary tree in 1837.

Genetic engineering

The fact that every living thing is on the same tree of life, and that the language of DNA is universal among all those life forms, means that DNA from a virus is the same as that from an elephant.

In the 1970s, scientists invented techniques to cut out DNA from one species and insert it into another – the cell doesn't care where the DNA comes from, as long as the code works. This was the birth of genetic engineering. At exactly the same time that popular music was being sampled and remixed for the first time, so was life.

Genetic engineering has changed all biology. We routinely shift genes from one species to another to do basic science, to manufacture drugs, to make genetically modified foods and to build therapies for diseases. Almost all of it is done in bacteria, but there are some more fanciful genetically modified animals, such as cats with fluorescent jellyfish genes that glow in the dark (useful for HIV research), and goats with genes from spiders so that they make spider silk in their udders when they produce milk (because spider silk is an extremely tough yet elastic material, and we can't farm spiders – they're often cannibals).

Traditionally, DNA could only be fused and mixed by individuals within the same species having sex. Unsurprisingly, a goat and a spider cannot have sex, but with genetic engineering we have bypassed this impossibility.

CRISPR/Cas9

Traditional forms of genetic engineering can be cumbersome and slow. Since 2012, a new technique has been gaining popularity which is quick and incredibly precise. It's called CRISPR/Cas9 (pronounced *crisper*), and was invented by Jennifer Doudna and Emmanuelle Charpentier, and modified by other scientists.

They identified a system that some bacteria have for cutting out bits of DNA from their own genomes, and adapted it so that they could snip out any bit of DNA from virtually any organism.

CRISPR is already being used in all sorts of organisms, from crops to monkeys to fruit flies, to do basic research, to aid conservation, for gene therapy and to improve drug design.

But it also means that we can tinker with the human genome quite easily. It's not legal to alter human DNA in embryos that could be born, but soon we may be able to correct faulty genes and eradicate diseases such as cystic fibrosis or muscular dystrophy not just from an individual who has the condition, but from the sperm and egg, so that they are erased permanently from all subsequent generations. As with all genetic engineering, the ethics of these inventions are now a core part of the science, and we as a society must work out whether we should do the things that we now can do.

Remixing biology

In the twenty-first century, it became possible to not just tinker with individual genes, but to build whole genetic circuits – a bit like electronic circuits. Scientists argued that electrical parts were standardized so they just fitted together without having to be redesigned; if they could make genes and other bits of DNA slot together, then anyone could build complex genetically engineered organisms that could be useful for specific jobs.

This is called synthetic biology, and is being used to make drugs, diesel, interesting materials such as spider silk, detectors for pollutants, and even tools to tell you when your meat is past its sell-by date. Nowadays schoolchildren can build genetic circuits using DNA from many different organisms with very little knowledge of genetics.

There's another branch of synthetic biology, in which scientists rewrite DNA itself. Instead of using the four natural letters of genetic code, they have begun to invent new letters that can be incorporated into DNA. There are new types of double-helix molecules that can reproduce like DNA. These might be used for designing new types of life, or for making drugs that won't be digested by our immune systems.

Only 150 years after the beginning of genetics, we are now capable of rewriting the whole of the code of life. And it's already taking us in some unexpected directions.

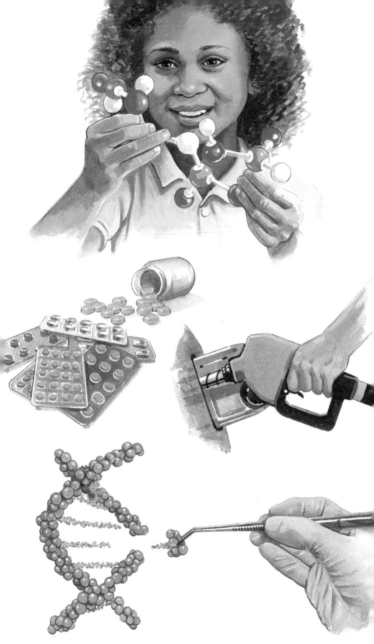

DNA as storage

DNA is data. That code is a digital format which stores information about the life form that inherited it. It's a remarkably stable format. Along with the Neanderthal DNA, we've managed to retrieve genomes from plants locked in ice cores for 800,000 years. And since we became adept at handling and manipulating DNA, the idea that we could encode information has become a reality.

In 2013, a team in Cambridge took digital files of Shakespeare's sonnets, a video of Martin Luther King's 'I have a dream' speech, and Crick and Watson's original 1953 paper on the structure of DNA, and converted them into a new DNA code. They synthesized the DNA, and dried it into a powder a bit like Friedrich Miescher's samples from the nineteenth century. They sent it to a lab in Germany, with instructions on how to decode it. The German team unscrambled it and translated it back into the original files, with an error rate of exactly zero.

Artificially writing code into DNA is slow, and it's also slow to decode. At the moment this technique is only of use for archive storage of data. But in the future we might be running computers not off microchips, but on DNA.

DNA is a simple language – an alphabet of just four letters, and twenty-one words. Yet it codes the most complex things that have ever existed – life. For four billion years, this code has been passed down from cell to cell, subtly changing to fuel evolution, and to determine the rules of inheritance. In the 150 years since DNA was first discovered, we are beginning to understand it, and even radically rewrite the code of life on Earth.

Further reading

Life's Greatest Secret by Matthew Cobb (Profile Books, 2015) is the most comprehensive telling of the story of twentieth-century genetics, and the cracking of the code.

The Double Helix by James Watson (Weidenfeld & Nicolson, 2010) is a classic, rollicking tale of the discovery of the double helix, though it is prone to myth-making and tall tales, and, in its original text, breathtaking sexism directed at Rosalind Franklin (Watson later acknowledged her contributions much more gracefully).

To bring us right up to date, *A Crack in Creation* by Jennifer Doudna (Houghton Mifflin, 2017) is a first-hand account of the invention of CRISPR/Cas9, by one of its inventors.

The Vital Question and *Life Ascending*, both by Nick Lane (Profile Books, 2015 and 2001), are brilliant analyses of many aspects of evolution and why life is the way it is, and both focus on the beginnings of life on Earth.

Two books by me cover many of the topics in greater depth. *Creation* (Viking, 2013) is in two halves: the first about the origin of life and the second about synthetic biology and genetic engineering. *A Brief History of Everyone Who Ever Lived* (Weidenfeld & Nicolson, 2016) re-examines the entirety of human history, using DNA as a source. It starts with the Neanderthals and ends with the future of humankind.